P9-DFH-957

POSTMODERN ENCOUNTERS

Thomas Kuhn and the Science Wars

Ziauddin Sardar

Series editor: Richard Appignanesi

ICON BOOKS UK

TOTEM BOOKS USA

Published in the UK in 2000
by Icon Books Ltd., Grange Road,
Duxford, Cambridge CB2 4QF
email: info@iconbooks.co.uk
www.iconbooks.co.uk

Published in the USA in 2000
by Totem Books
Inquiries to: PO Box 223,
Canal Street Station,
New York, NY 10013

Distributed in the UK, Europe,
Canada, South Africa and Asia
by the Penguin Group:
Penguin Books Ltd.,
27 Wrights Lane,
London W8 5TZ

In the United States,
distributed to the trade by
National Book Network Inc.,
4720 Boston Way, Lanham,
Maryland 20706

Published in Australia in 2000
by Allen & Unwin Pty. Ltd.,
PO Box 8500, 9 Atchison Street,
St. Leonards, NSW 2065

Library of Congress catalog
card number applied for

ISBN 1 84046 136 5

Typesetting by Wayzgoose

Printed and bound in the UK by
Cox & Wyman Ltd., Reading

A Postmodern Trick: Background to the Sokal Affair

In early 1995, the editors of the journal *Social Text* were putting the finishing touches to a special issue on 'Science Wars'. *Social Text*, edited at the Center for the Critical Analysis of Contemporary Culture, Rutgers University, and published by the Duke University Press, is considered by many to be a primary journal of critical theory. The double issue on 'Science Wars' was being produced in the wake of, and as a response to, a number of recent attempts to mount an aggressive defence of science and to question the very integrity of the 'cultural studies' approach and criticism of science.

The publication of *The Bell Curve* (1994) by Richard Herrnstein and Charles Murray had rekindled the old IQ and eugenics controversy. An exhibition on 'Science in American Life' at the Smithsonian's Museum of American History was strongly attacked by various professional scientific societies for presenting a less-than-favourable image of science. A series of lavish, well-funded and highly publicised conferences had mobilised a broad coalition of scientists, social scientists and other scholars for the defence of science. The most publicised and effective of these was 'The Flight from Science and

Reason' conference, sponsored by the New York Academy of Sciences, and held in New York during the summer of 1995.[1] The conference declared that there was a real threat to science from sociologists, historians, philosophers and feminists who work in the field of 'science and technology studies' (STS). It attacked the social theories of science, declared feminist epistemology (theory of knowledge) a 'dead horse' and the criticism of science a 'common nonsense', and described most critics of science as 'charlatans'.

But it was Paul Gross and Norman Levitt's *Higher Superstition: The Academic Left and Its Quarrels with Science* (1994), published a year earlier, that more than anything else motivated the *Social Text* issue on 'Science Wars'. Biologist Gross and mathematician Levitt declared that the 'academic left', 'a large and influential segment of the American academic community', was intrinsically anti-science. This anti-science hostility is based not just on the academic left's dislike of the uses to which science and technology are put by political and economic forces – such as military hardware, surveillance, industrial pollution and destruction of the environment. Even scientists regret these abuses of science and technology. The hostility 'extends to

the social structures through which science is insti-
tutionalised, and to a mentality that is taken, rightly
or wrongly, as characteristic of scientists. More
surprisingly, there is open hostility toward the
actual content of scientific knowledge and toward the
assumption, which one might have supposed universal
among educated people, that scientific knowledge is
reasonably reliable and rests on a sound methodology'.[2] Gross and Levitt dubbed this hostility
'medieval', a clear rejection of 'the strongest heritage
of the Enlightenment' and a denial of 'progress'.

The purpose of the 'Science Wars' issue was to
answer 'the shrill tone of backlash' against feminist,
multiculturalist, and social critics of science. *Social
Text* saw this as nothing less than scare stories about
an anti-science movement – a backlash designed to
intimidate anyone who dares to question the gender-
laden assumptions of science, the capitalist foundations of scientific empiricism and the destructive
effects of science and technology on society and
environment. 'This is not the first time that a hue
and cry has been raised about the decline of science's
authority, or its imperilment on all sides by the
forces of irrationalism', wrote Andrew Ross,
Director of the American Studies programme at
New York University, and a guru of the critical

science movement, in his introduction to the issue.[3] Science has become a new religion, asserted Ross, and the questioning of science is now perceived as a clear and present danger to civilisation. Given that science has now been completely compromised, industrialised and commodified, 'the militant resurgence of belief in its pristine truth claims is not hard to understand. But the crusaders behind the Science Wars are not about to throw the money-lenders out of the temple. Their wrath is aimed, above all, at those who show how the temple was built and how its rituals are maintained – the con-structionist academic left'.[4] *Social Text* had gathered many of these 'constructionists' – including Sandra Harding, *doyenne* of feminist epistemology; Steve Fuller, the founder of social epistemology; and Hilary Rose, *grande dame* of the British radical science movement – to answer the charges thrown at them by the defenders of science. (See 'Key Ideas' at the end of this book for an explanation of 'constructionism'.) There was, however, one late addition.

Just as work on the journal was being concluded, a new paper was submitted for specific publication in the 'Science Wars' issue. Entitled 'Transgressing the boundaries: towards a transformative hermeneutics of quantum gravity', it was written by Alan D. Sokal, a professor of physics at New York University. The paper duly appeared in the Spring/Summer 1996 issue of *Social Text*.[5]

A reasonably critical examination of Sokal's paper would easily have aroused the suspicions of the editors. The paper purports to argue that unifying the currently incompatible theories of quantum mechanics and general relativity would produce a postmodern, 'liberatory' science. It contains some deliciously daft assertions. For example, it suggests that pi, far from being a constant and universal, is actually relative to the position of an observer and is thus subject to 'ineluctable historicity'. The relativism that the paper presents crosses the boundaries of lunacy. The bibliography appended to the paper clearly reads like a 'who's who' of science critics, and bears little relationship to the contents of the paper. By cleverly cloaking his absurdities with quotations from postmodern masters such as Derrida, Lyotard and Lacan, and more flattering citations from the editors of *Social Text*, Andrew Ross and Stanley Aronowitz, Sokal was able to get away with his parody.

Sokal promptly revealed his hoax in the pages of *Lingua Franca*.[6] The scandal hit the front page of the *New York Times*, *International Herald Tribune*, *Le Monde* and numerous other newspapers. The Science Wars became public. This is the latest salvo in the long war between scientists and their critics. It is a fierce war that has taken many casualties – both in terms of loss of careers and marginalisation of ideas – and affects the very soul of science. In the first instance, we can identify the 1960s and the publication of Thomas S. Kuhn's *The Structure of Scientific Revolutions* (1962) as the flashpoint at which open warfare emerged between defenders and critics of science. Kuhn showed that far from being the pursuit of objectivity and truth, science was little more than problem-solving within accepted patterns of beliefs. But Kuhn's ideas did not emerge ready-made; they have a historical context. Before it became a full-blown war, the hostility between scientists and critics simmered just below the surface for decades. We can trace the Science Wars back to the formation of the radical science movement.

The Origins of Science Wars

For nearly a century, up to the First World War, science had been almost universally seen in heroic terms. The lone scientists struggled against all odds for the sake of Truth. Science was a pure, autonomous activity, separate from technology and industry and above society. The purity of scientific research was particularly ensured in universities where research was pursued for the sake of Knowledge and future generations of scientists were trained. The term 'scientist' was coined in the 1830s by William Whewell (1794–1866), a physicist and historian of science; before that scientists were seen as 'natural philosophers'. Whewell saw the scientist as someone engaged in a unique social role who required protection and had autonomous existence from the rest of the society. The scientists, particularly Great Scientists, were the object of inquiry for historians and philosophers of science. Thus the emphasis of the history of science was on discoveries by great scientists and the justification of these discoveries in the unique objectivity, impartiality and universality of science. But while science was true at all or any time, there were 'errors' to be corrected. The explanation in the history books was that scientists who persisted in error once the truth was shown were somehow bad scientists.

However, not everyone was totally convinced about the absolute purity of science. A few scientists expressed concern over the future direction of science. In particular, Ernst Mach (1838–1916), physicist and philosopher of science, argued, in the decades preceding the First World War when experimental physical scientists were becoming involved with industry and the military, for a more scaled-down, user-friendly science directed towards meeting human ends. Mach was opposed by Max Planck (1858–1947), another eminent German physicist, who supported a more autonomous ideal of science. Their debate spanned many important epistemological and political issues, the most lasting of which saw realism (the idea that given statements are without exception made true or false by mind-independent facts) set against instrumentalism (the theory that we cannot make claims about truth or reality beyond whatever is borne out by empirical evidence) in the philosophy of science. But Planck's victory over Mach, widely recognised, further entrenched the acceptance of a Platonic vision of the experimental scientist guided solely by an over-arching commitment to truth.

But the First World War seriously dented this mythological belief in the purity of science. It exposed the technological weaknesses of the British Empire and led to direct government intervention in the management of science. The monopoly of universities as research institutions was broken as new institutions were established with public and private funding. To many intellectuals and scholars, particularly of Marxist persuasion, a relationship between science and economics became plainly evident. It led to the formation, in 1918, of the National Union of Scientific Workers (later Association of Scientific Workers) with a categorically socialist agenda for science. Increased expenditure on science, along with centralised planning, it was argued, would release the liberating potentials of science.

The connection between science and ideology was made explicit in 1931 when a conference on the history of science in London played host to a delegation from the (then) Soviet Union. The key event at the conference was a paper by Boris Hessen on 'The Social and Economic Roots of Newton's *Principia*'.[7] Hessen argued that Newton's major work was not so much a product of scientific genius or a result of the internal logic of science, but rather a consequence of social and economic forces in seventeenth-century Britain. It fulfilled the needs of the British bourgeoisie. The young British left-wing scientists and historians of science attending the conference took a few years to grasp the full import of Hessen's arguments. But with the publication of

J.D. Bernal's *The Social Function of Science* in 1939, the radical science movement had truly arrived. Bernal saw science as a natural ally of socialism: its function was to serve the people and liberate them from capitalism. Bernal combined his Marxist humanitarianism with technocratic and reductionist motives. Despite all its problems, Bernal held on to his faith in science as an objective, neutral mode of inquiry that could produce peace and plenty for all, were it not for the corruption of science under capitalism.

The idea of a 'socialist science', first suggested in the Soviet Union, also gained currency in Britain. But its realisation in the Soviet Union came to be seen as a crude and opportunistic exercise. The Lysenko affair of the 1940s and 50s, which involved Soviet geneticists arguing that heredity can be transformed by means of environmental manipulation and grafting, did great harm to the idea of a socialist science. Later, the avowedly radical British Society for Social Responsibility in Science did manage to organise a conference on the question: 'Is there a socialist science?', but the outcome remained definitely inconclusive.

In the popular perception of science, the Second World War completed what the First World War had started. This time, science was seen to be running the show on the battlefield, as well as moving into government. Scientists were responsible not just for developing new and deadlier forms of chemical and biological weapons, but for conceiving, producing and finally unleashing The Bomb. The mushroom clouds of the atomic weapons dropped on Hiroshima and Nagasaki declared the end of the era of scientific innocence. Now the connection between science and war was all too evident, the complicit relationship between science and politics had come to the fore, and all notions of scientific autonomy had evaporated. The public, which had hitherto concerned itself largely with the benefits of science, suddenly became concerned with its devastating downside.

The protest against militarised science, starting with the launching of the dissident journal *Bulletin of the Atomic Scientists* by nuclear physicists totally disenchanted by the Manhattan Project in the US, was consolidated by the emergence of CND (Campaign for Nuclear Disarmament) in the later 1950s. The first salvo of what we now see as Science Wars was fired during this period. Many scientists were concerned that the Bomb should not be seen as an inescapable consequence of physics.

This would discourage bright young scientists with ethical concerns about the Bomb from pursuing a career in physics. The ploy was to claim that science is neutral: it is neither good nor bad; it is society that puts it to good or bad use. The neutrality argument became a dominant defence of science during the 1950s and 60s; and it enabled many scientists to work in atomic physics, even accepting grants from defence establishments, while professing to be politically radical.

While the radical science movement questioned the neutrality of science, debated its social function, and campaigned against the Bomb, it still saw science very much as a Western concern. One of the great myths about science is that it is largely a product of Western civilisation, with little or no contribution from other cultures and civilisations. Indeed, Western histories of science made sciences of all non-Western cultures totally invisible. Thus, while radical historians of science were eager to show how social forces shaped the development of science, they remained largely silent on the role non-Western cultures played in shaping science. So Bernal, for example, explains his reasons for writing *Science in History* as follows:

In the last thirty years, largely owing to the impact of Marxist thought, the idea has grown that not only the means used by natural scientists in their researches but also the very guiding ideas of their theoretical approach are conditioned by the events and pressures of society. This idea has been violently opposed and as energetically supported; but in the controversy the earlier view of the direct impact of science on society has become overshadowed. It is my purpose to emphasise once more to what extent the advance of natural science has helped to determine that of society itself; not only in the economic changes brought about by the application of scientific discoveries, but also by the effect on the general frame of thought of the impact of new scientific discoveries.[8]

But Bernal perceives 'society' largely as an autonomous Western society; and civilisation (always in the singular) for him is essentially Western civilisation which starts with the Greeks and progresses linearly to 'our time'. In a four-volume study, he devotes less than ten pages to Islamic science. China and India do not even get a mention.

But the historiography of science was about to change quite radically. The foundation for this change had been established by two truly monumental studies. The first was George Sarton's *Introduction to the History of Science*, published between 1927 and 1948. What is surprising about Sarton's study is that the first three volumes of the four-volume chronological study are largely devoted to science in Islam. Sarton not only makes it clear that Western science is inconceivable without Islam, but suggests that the sheer scientific contribution of Islam, both in terms of quantity and quality, should concern those who see science purely as a Western enterprise. The second study, Joseph Needham's massive multi-volume *Science and Civilisation in China*, published from 1954 onwards, performed the same function for China, even more eloquently.

Both Sarton and Needham showed that science was not limited to Western societies; non-Western societies also had highly developed and sophisticated cultures of science. But for them this did not mean that there could be different sciences; or indeed different definitions of science. Needham was able to isolate the 'problematic' bits of Chinese science, such as acupuncture, and label them as 'non-science'. Science may be tainted with politics, and embroiled with the military, but it was still largely seen as a neutral, objective, universal pursuit of truth.

In the United States, the 1950s, when Kuhn was going through university, was a period of extreme political repression. There was a real reign of terror in American universities; people could be sacked and victimised, with no redress, when some 'Un-American Activities Committee' put pressure on their employers. We have heard of many Hollywood victims of Senator McCarthy's anti-American trials; but the academics who suffered were more numerous and equally important. The use of the very word 'social' was seen to connote 'socialist', which was equivalent to 'Commie'. The few scholars who had been promoting a social history of science were unheard; others were threatened. The leading historian of science, Alexandre Koyré, went so far in his idealised representation of Galileo as to deny not just a social context to Galileo's work, but even to doubt that he did his famous experiments. Anything that remotely suggested a social influence on science or scientists, Koyré aggressively dismissed as 'Marxist'. Due to Koyré's influence, Bernal's 'externalist' approach to the history of science remained firmly outside the corridors of academia.

The physicists, seeing that their prestige was waning thanks to the Bomb, decided to become the self-appointed defenders of scientific rectitude. The object of their scorn was not scholarly criticism of science – for there was hardly any in the US – but irrational forces such as creationism. A crisis in 'big science', with the cancellation of the big high-energy projects, forced them to conjure up an instant enemy. Fortunately, one was at hand. The publication of *The Velikovsky Affair* (1968) by Alfred De Grazia generated a heated controversy. Immanuel Velikovsky was a psychoanalyst who had mastered many languages and burrowed through mountains of historical sources in an attempt to prove his bizarre thesis of violent changes in the orbits of the planets in Biblical times, arguing for the priority of the Biblical Hebrew civilisation. In his *Worlds in Collision* (1950) and other books, he also predicted various properties of the nearby planets and inner space, which were later confirmed to an embarrassing degree. The response of the US astrophysical community was fierce, some said hysterical. De Grazia showed that the scientists broke all their own rules in their treatment of Velikovsky and the examination of his claims. That mini skirmish in the Science War only subsided when, eventually, Carl Sagan debated personally with Velikovsky at a meeting of the American Association for the Advancement of Science – and it became suddenly clear to everyone that the old Velikovsky, however brilliant and prescient he may have been perceived by some, was really a crank. But the Velikovsky affair, and others of a similar sort, performed an important cultural function: by demonstrating that they can beat their enemies so thoroughly, the scientists hoped to continue to enjoy public and government support. All this is to show that the neglect of the social dimensions of science in the USA was not the result of considered judgements of scholars, but a consequence of the Cold War.

While things were also tough for radicals in Britain during the 1950s, it was still possible for Joseph Needham to praise Mao Tse-tung in the first volume of his *Science and Civilisation in China*. Although he was ostracised to a considerable degree for his support of the North Korean claims of germ warfare by the Americans, by the end of the decade he was ensconced as Master of Caius College, Cambridge. It was thus still possible in Britain to be an avowed radical and get away with it. By the end of the 1960s, the radical science movement in Britain had become quite prominent; although only a few radicals actually engaged in scholarly work.

Among these were Hilary and Steven Rose, who were engaged in producing a Marxist synthesis of 'science and society'; Robert Young, later to become the guru of the radical science movement and resident mentor of the *Radical Science Journal*, who was creating a social interpretation of Darwin and Darwinism (which outraged the Darwin hagiographical industry); and Jerry Ravetz, whose work outside the history of science was concerned largely with the analysis of nuclear insanity and with corruption within 'big science'.

It was in this Cold War atmosphere that Kuhn produced his seminal work.

Kuhn and *The Structure of Scientific Revolutions*

With the publication of *The Structure of Scientific Revolutions* in 1962, Thomas Samuel Kuhn inaugurated a new epoch in the understanding of science. Born in Cincinnati, Ohio, in 1922, Kuhn studied physics at Harvard University. He went on to do graduate studies in theoretical physics, but decided to change to history of science just before finishing his dissertation. As he describes it in his preface to *Structure*:

[A] fortunate involvement with an experimental college course treating physical science for the non-scientist provided my first exposure to out-of-date scientific theory and practice and radically undermined some of my basic conceptions about the nature of science and the reasons for its special success. These conceptions were ones I had previously drawn partly from scientific training itself and partly from a long-standing avocational interest in the philosophy of science. Somehow, whatever their pedagogic utility and their abstract plausibility, those notions did not at all fit the enterprise that historical study displayed. Yet they were and are fundamental to any discussion of science, and their failure of verisimilitude therefore seemed thoroughly worth pursuing. The result was a drastic shift from physics to history of science, and then, gradually, from relatively straightforward historical problems back to the more philosophical concerns that initially led me to history.[9]

During the period that Kuhn studied, thought and wrote *Structure*, Harvard was steeped in a particular ideology. Cambridge, Massachusetts was the hub of the scientists who created Big Science and worked on the Manhattan Project. And they were bringing that ideology, as well as science on an industrial scale, back to the campuses. In particular, the president of Harvard, James Bryant Conant, had been instrumental in bringing the German large-scale 'industrial' model of scientific research to American academia after the First World War. Conant was also the US atomic bomb administrator, mediating between Congress and the Los Alamos team, and he was the person who convinced President Truman to argue that dropping the atom bomb on Hiroshima was 'inevitable'. Conant became Kuhn's mentor and was responsible for persuading him to teach in the General Education in Science programme, where he honed the theses of *Structure*, which is dedicated to Conant.

Thus Kuhn grew up in a science that was becoming industrialised and had been transformed into a career rather than a vocation. The dominant school of philosophy of science during that period was largely the product of the work of the Vienna Circle and Karl Popper (1902–94). The Circle, consisting of philosophers, mathematicians and scientists, was established in the 1920s. Its intellectual ancestry began with Ernst Mach, and some influential members, such as Rudolf Carnap and Otto Neurath, saw it as a means of advancing anti-clerical and socialist ideas. While Popper was loosely associated with the Circle, he was not a member of it, being highly critical of some aspects of its philosophical position. The Circle asserted that metaphysics and theology were meaningless, for they consisted of propositions that could not be verified. Its own doctrine, known as logical positivism, conceived philosophy as purely analytical, based on formal logic, and the only legitimate component of scientific discourse. After the assassination of a member and Hitler's invasion of Austria, the members of the Circle emigrated to the US.

Popper found a job in New Zealand, where he stayed until coming as Professor to the London School of Economics in 1948, on the strength of his anti-Communist opus, *The Open Society and its Enemies* (1945). Popper's ideas on the nature of scientific procedure were developed in *The Logic of Scientific Discovery* (German original 1934; translation 1959). He disagreed with traditional beliefs about 'induction', arguing that no number of particular instances, for example of A being a B, can establish a universal principle that all A's are B's. He also disagreed with the reliance placed on 'verification' by the Vienna Circle. For him 'falsification', or rather 'falsifiability', was the genuine demarcation between science and non-science. He therefore argued that there is no final truth in science, and that scientific progress is achieved by 'conjectures and refutations' (this became the title of his book of essays, published in 1963). And for Popper, the self-critical spirit is the essence of science.

Popper and the original Vienna Circle had a common commitment to philosophy of science as relevant to societal, even political concerns. By contrast, Kuhn's mentors were academics and technicians. He did his doctorate under P.W. Bridgman, who was quite a distinguished amateur philosopher of science. But Bridgman was not a 'professing' philosopher, and his scientific work was rigorously, ruthlessly, mind-numbingly practical: the achievement of very high pressures. So Kuhn matured during a period when the study of science by scholars was seen in strictly scholastic terms. Even when Conant sponsored scholarly work with his 'Harvard Case Studies in the History of Science', which subsequently became highly influential in science studies, it was strictly 'internal'. The struggles of the Vienna Positivists were converted into a crabbed, restrictive doctrine by their American disciples. In the States, the war of Science with Theology was over (except for the creationists, who were out in populist Middle America). The ideology of science that

Kuhn imbibed was not a fighting creed. All that he experienced was an exclusive 'presentist' doctrine that contemporary science is the arbiter by which all other productions are judged, be they creative work, belief, or even the science of the past.

After completing his Ph.D., Kuhn remained at Harvard as a Junior Fellow, but he left when a job in History of Science went, not to himself, but to the then more established historian, I. Bernard Cohen. (The committee that denied tenure to Kuhn at Harvard in 1956 regarded him as beholden to Conant – who had since left the Harvard presidency to become the first US ambassador to West Germany.) He went to teach at the University of California, Berkeley, where he did his most productive work. Then he moved to the Institute for Advanced Study in Princeton, and finally returned to Cambridge, Massachusetts, but this time to the Massachusetts Institute of Technology. His earlier research focused on the history of thermodynamics, and his first book, *The Copernican Revolution* (1957), with a foreword by Conant, is a study of the development of heliocentrism during the Renaissance. But it was *Structure*, seen by many as one of the key books of the twentieth century, that established his reputation.

Kuhn looks at science from the particular perspective of a professional historian. He explores bigger themes, such as what is science really like in its actual practice, with concrete, empirical analysis. In *Structure*, he argues that scientists are not bold adventurers discovering new truths; rather they are puzzle-solvers working within an established worldview. Kuhn used the term 'paradigm' to describe the belief system that underpins puzzle-solving in science. By using the term 'paradigm', he writes, 'I mean to suggest some accepted examples of actual scientific practice – examples which include law, theory, application, and instrumentation together – provide models from which spring particular coherent traditions of scientific research. These are traditions which history describes under such rubrics as "Ptolemaic Astronomy" (or "Copernican"), "Aristotelian dynamics" (or "Newtonian"), "corpuscular optics" (or "wave optics") and so on'.[10] The term 'paradigm' is closely related to 'normal science': those who work within a dogmatic, shared paradigm use its resources to refine theories, explain puzzling data, establish increasingly precise measures of standards, and do other necessary work to expand the boundaries of normal science.

In Kuhn's scheme, this dogmatic stability is

punctuated by occasional revolutions. He describes the onset of revolutionary science in vivid terms: 'Normal science . . . often suppresses fundamental novelties because they are necessarily subversive of its basic commitments . . . [but] when the profession can no longer evade anomalies that subvert the existing tradition of scientific practice . . .',[11] then extraordinary investigations begin. A point is reached when the crisis can only be solved by revolution in which the old paradigm gives way to the formulation of a new paradigm. Thus 'revolutionary science' takes over; but what was once revolutionary itself settles down to become the new orthodoxy: the new normal science. So science progresses, argues Kuhn, through cycles: normal science followed by revolution followed again by normal science and then again by revolution. Each paradigm may produce a particular work that defines and shapes the paradigm: Aristotle's *Physics*, Newton's *Principia* and *Opticks* and Lyell's *Geology* are examples of works that defined the paradigms of particular branches of science at particular times.

In sharp contrast to the traditional picture of science as a progressive, gradual, cumulative acquisition of knowledge based on rationally chosen experimental frameworks, Kuhn presented 'normal' science as a dogmatic enterprise. If we regard out-moded scientific theories such as Aristotelian dynamics, phlogistic chemistry, or caloric thermo-dynamics as myths, he argues, than we can just as logically consider the current theories to be irra-tional and dogmatic:

If these out-of-date beliefs are to be called myths, then myths can be produced by the same sorts of methods and held for the same sorts of reasons that now lead to scientific knowledge. If, on the other hand, they are to be called science, then science has included bodies of belief quite incompatible with the ones we hold today. . . . [This] makes it difficult to see scientific development as a process of accretion.[12]

Throughout the book he uses historical examples to throw light on the current practice, identifying com-mon factors and emphasising the flawed nature of the scientific method. Thus, the scientific method – the idealised process of observation, experimenta-tion, deduction and conclusion – on which much of science's claims to objectivity and universalism are based, turns out to be a mirage. Kuhn suggests that it is the paradigm that determines the kinds of experiments scientists perform, the types of questions they ask, and the problems they consider important. Without a given paradigm, scientists cannot even gather 'facts': 'in the absence of paradigm or some candidate for paradigm, all of the facts that could possibly pertain to the development of a given science are likely to seem equally relevant. As a result, early fact-gathering is a far more nearly random activity than the one that subsequent scientific development makes familiar'.[13] A shift in the paradigm alters the fundamental concepts underlying research and inspires new standards of evidence, new research techniques, and new pathways of theory and experiment that are radically 'incommensurate' with the old ones.

Most scientific activity, according to Kuhn, takes place under the rubric of 'normal science', which is the science we find in textbooks, and which requires that research be 'based upon one or more past scientific achievements, achievements that some particular scientific community acknowledges for a time as supplying the foundation for its further practice'.[14] This restrictive, closed puzzle-solving science has both its advantages and disadvantages. On one hand, it enables the scientific community to gather data on a systematic basis and rapidly push forward the frontiers of science:

When the individual scientist can take a paradigm for granted, he need no longer, in his major works, attempt to build his field anew, starting from first principles and justifying each concept introduced. That can be left to the writer of textbooks. Given a textbook, however, the creative scientist can begin his research where it leaves off and thus concentrate exclusively upon the subtlest and most esoteric aspects of the natural phenomena that concern his group.[15]

On the other hand, normal science isolates the scientific community from the outside. Socially important problems that cannot be reduced to puzzle-solving form, Kuhn suggests, are ignored, and anything outside the conceptual and instrumental scope of the paradigm is seen as irrelevant.

Kuhn's approach to science was essentially a reaction to the 'Whig interpretation of history', that history is a linear progress of increasing liberty culminating in the present. Whig history reads the past backwards and explains the present as a cumulative product of past achievements. The debunking of Whig history in the history of science was started, amongst others, by Alexandre Koyré, to whom Kuhn acknowledges a major intellectual debt. Kuhn realised that to appreciate how a historical tradition develops, one has to understand the social behaviour of those involved in shaping the tradition. 'It is this insight', writes Barry Barnes, 'combined with his historical sensibility, which gives Kuhn's work its originality and significance. The continuation of a form of culture implies mechanisms of socialisation and knowledge transmission, procedures for displaying the range of accepted meanings and representation, methods of ratifying acceptable innovations and giving them stamp of legitimacy. All of these must be kept operative by the members of the culture themselves, if its concepts and representations are to be kept in existence. Where there is a continuing form of culture there must be sources of cognitive authority and control.'[16] Kuhn represents scientific research as a product of a complex interaction between a research community, its authoritative tradition, and its environment. Nowhere in the entire research process are 'reason' and 'logic' the sole criteria for advances in scientific knowledge.

When it was first published, *Structure* generated a great deal of controversy. The reaction from scientists was not surprising: after all, Kuhn had pulled the carpet from underneath the perceived notion of the scientist as a heroic, open-minded, disinterested seeker of Truth and interrogator of nature and reality. And he had reduced, as parodies of his account suggested, science to nothing more than long periods of boring conformist activity punctuated by outbreaks of irrational deviance. But the philosophers of science too were hostile to Kuhn, for they had been, up to now, responsible for producing accounts of the nature of scientific research and progress. Kuhn's account was barely recognisable next to their formalised and idealised product. Comparisons with theology, religious conversions and political revolutions horrified both scientists and philosophers of science. The philosophers also found Kuhn's relativism quite repulsive. In history and philosophy of science (HPS) circles, *Structure* was described as unoriginal, dry and confused. Stephen Toulmin had already floated an idea of 'frameworks' as a counter to the positivistic fact-collecting image; and the historian-philosopher R. G. Collingwood had previously introduced similar ideas. In particular, the Harvard pragmatist C. I. Lewis anticipated many of Kuhn's most radical statements concerning the incommensurability of world-views.

However, by the late 1960s, *Structure* began to be

accepted as a revolutionary work in the philosophy of science. It sold over a million copies in 20 languages, becoming one of the most influential academic books of the twentieth century. Its concept of paradigm shifts began to be used in such disciplines as political science and economics. In sociology, it was embraced wholeheartedly. Soon, a new discipline was to emerge: the critical sociology of science. According to the historian of science Ian Hacking, *Structure* spelled the end of the following notions:

Realism: that science is an attempt to find out about one real world; that truths about the world are true regardless of what people think; that the truth of science reflects some aspect of reality.

Demarcation: that there is a sharp distinction between scientific theories and other kinds of belief systems.

Cumulation: that science is cumulative and builds on what is already known, Einstein being a generalisation of Newton.

Observer-theory distinction: that there is a fairly sharp contrast between reports of observation and statements of theory.

Foundations: that observation and experiment

provide the foundations for and justification of hypothesis and theories.

Deductive structure of theories: that tests of theories proceed by deducing observation-reports from theoretical postulates.

Precision: that scientific concepts are rather precise and the terms used in science have fixed meanings.

Discovery and justification: that there are separate contexts of discovery and justification, and that we should distinguish the psychological or social circumstances in which a discovery is made, from the logical basis for justifying belief in facts that have been discovered.

The unity of science: that there should be one science about the one real world; less profound sciences are reducible to more profound ones: psychology is reducible to biology, biology to chemistry, chemistry to physics.[17]

Post-Kuhnian Developments

Kuhn represents a new phase in the ideology of science. He was immediately recognised as important by Popper and his group. They were engaged in an ideological struggle of their own, defending their version of the rationality of science. So the Popper group organised, in July 1965, an International

Colloquium in the Philosophy of Science – it was backed by a whole range of institutions including the British Society for the Philosophy of Science, London School of Economics and the International Union of History and Philosophy of Science – to undermine Kuhn. The purpose of the Colloquium was to bring Kuhn up against the combined might and criticism of the British philosophers. Among other things, 23 possible meanings of 'paradigm' were presented; and Popper identified Kuhn's 'normal science' as an enemy of science and civilisation. All the debates, and Kuhn's replies, were eventually published in *Criticism and the Growth of Knowledge* (1970).

Present at the Colloquium were two philosophers of science who would soon engage in their own mini science war: Imre Lakatos (1922–74) and Paul Feyerabend (1924–94). Lakatos was a Hungarian who fought in the anti-Nazi resistance and became prominent in the Communist government. But he was imprisoned and tortured by the Stalinists, and escaped during the 1956 uprising. He moved to London and joined the Popper group at the LSE. His best work, *Proofs and Refutations* (1976) (establishing a dialectical and historicist philosophy of mathematical proof) has a title that echoes

Popper; but a deeper influence is that of the Hungarian Marxist philosopher Georg Lukács and through him, Hegel. During the 1960s, Lakatos attempted to articulate a philosophy of scientific progress, 'The Methodology of Scientific Research Programmes', that combined Popper's idealism (a commitment to some sort of rational Method) with Kuhn's flat realism (scientists as mere puzzle-solvers).

Feyerabend was an Austrian who had had a varied career (including a spell in the army and a period of working with the Communist playwright Bertolt Brecht) before coming to the LSE. He debated brilliantly on behalf of Popper, but eventually developed drastically different ideas about science. These arose from his experiences at the University of California, Berkeley, in the 1960s, combining anti-war and radical protests with alternative medicine. Adopting an anarchist stance against what he saw as a reactionary hegemony in science, Feyerabend published his masterwork, *Against Method: Outline of an Anarchistic Theory of Knowledge*, almost a decade after the London Colloquium, in 1975. There he showed that any principle of Scientific Method has been violated by some great scientist, with Galileo as the cardinal sinner. This was presented as his side of a friendly debate with Lakatos, who tragically had already died in 1974. As a sort of tactical Anarchist, he held classes in Berkeley where he famously brought creationists, Darwinists, witches and other 'truth peddlers' to defend their opinions in front of the students.

Neither the pro- nor anti-Kuhn groups of the 1960s noticed that another book, published just after Kuhn's, had changed the debate forever. This was Rachel Carson's *Silent Spring* (1965), in which the harm to the environment caused by science-based technology was vividly displayed to the public. It became clear that when we consider who pays the wages of Kuhn's puzzle-solvers, who decides what problems they study, who controls the publication of their results, Kuhn's academic 'normal science' has been replaced by corporate, industrialised science. A few years later, in one of the most original works to be published after Kuhn, Jerry Ravetz, a philosopher and historian of science loosely attached to the radical science movement in Britain, argued that industrialised science was deeply vulnerable to corruption. The title of Ravetz's book, *Scientific Knowledge and Its Social Problems* (1971; 1996), was bold for its time. In the aftermath of Kuhn and Carson, the idea was gaining ground that science was a social activity that could lead to ethical problems. But to argue that knowledge itself could have social problems seemed both daring and illogical. To overcome this contradiction, Ravetz suggested that we need to abandon the idea that 'science discovers facts', or is 'true or false', or that knowledge is an automatic outcome of research. Rather, genuine scientific knowledge is the product of a lengthy social process, of which the major part occurs long after the research is completed. This means that science, interpreted as research or scholarship in the broadest sense, should be seen as 'craft work'. If science is seen as craft, then 'truth' is replaced by the idea of 'quality' in the evaluation of scientific output. Quality firmly places both the social and ethical aspects of science, as well as scientific uncertainty, on the agenda. Ravetz showed that in the overall practice of contemporary science one could identify four categories that were seriously problematic: shoddy science, entrepreneurial science (where securing grants is the name of the game), reckless science, and dirty science; and they are all involved with runaway technology. He showed further that quality in science depended largely on the morale and commitment of working scientists and was reinforced by the moral acumen of the leadership of the scientific communities. Now that the old idealism of 'little science' had lost its social and ideological foundation, and evaporated, a corresponding idealism was needed for industrialised 'big science'. 'Without such an idealism, science would be very vulnerable to corruption, leading to universal rule by mediocrity or worse.'[18]

Ravetz's concern with quality found little resonance with the history and philosophy of science (HPS) community. This was partly because quality as a problem was not seen by philosophers as their concern; and partly because HPS was too concerned with erecting disciplinary boundaries in the academic marketplace. Reflective professional scientists found that HPS people were even more dismissive of their work. Thus John Ziman's *Public Knowledge* (1968), an original meditation by a working scientist on the social character of science, received zero interest from the HPS community. This was all rather ironic, as the whole HPS effort got its big start as a 'bridge subject', in a reaction partly to the Bomb and partly to C.P. Snow's suggestion that science and the humanities were divided into two watertight cultures.

Ravetz devotes considerable space in *Scientific Knowledge and Its Social Problems* to the idea of scientific 'fact'. He shows how the results of research go through social processes of testing, eventually to become 'facts' and sometimes even 'knowledge'. He explains how it can be that different courses teach radically different versions of the same fact, how different editions of the same textbook move silently from one unquestionable version to another, and how finally students learn some particular vulgarised version of the fact as absolute truth. The examination of how the results of scientific research come to be seen as facts became a field of inquiry in its own right towards the end of the 1970s. Up to then, the sociologists had done their part in the general ideological programme whereby philosophers of science proved that science always got it right and historians showed that it happened. In the pre-war period, sociologists had nothing but adulation for science – as can be seen in the work of R.K. Merton, commonly regarded as an early pioneer in the social studies of science. But Kuhn changed all that. Sociologists responded enthusiastically to Kuhn. This was partly because he provided them with a way to emulate science itself: by creating its own paradigm, sociology could become a science like physics. And partly it opened up a whole new avenue of sociological inquiry. Sociology of Science, also dubbed Sociology of Knowledge, became a new, rapidly expanding discipline. Even anthropologists got in on the act: they studied scientists as exotic tribes with their own norms and rituals. Essentially, what both sociologists and anthropologists try to show is that scientific facts, as Ravetz had already suggested, are not 'discovered'; rather every fact has a socio-technical history associated with it. Among these scholars, the general term 'construction' was substituted for 'discovery'; the question left open was the degree to which this 'construction' is constrained by some objective reality 'out there'.

The constructionist studies of science, such as anthropologist Karin Knorr-Cetina's *The Manufacture of Knowledge* (1981) or the more recent *The Golem* (1993) by sociologists Harry Collins and Trevor Pinch, have two basic aims. They intend to show, first, that industrialised science manufactures both the 'facts' of science and the 'truth' they are supposed to express. And, second, the scientific method itself, far from being a paradigm of non-local universality, is nothing more than opportunistic logic and a locally situated form of practice that is rooted in local social action. Undoubtedly, the most famous constructionist study is *Laboratory Life: Social Construction of*

Scientific Facts (1979; 1986) by Bruno Latour and Steve Woolgar. This duo of French and British sociologists examined the detailed history of a single fact: the existence of Thyrotropin Releasing Factor (Hormone), or TRF(H) for short. TRF(H) first emerged in 1962 with the statement that 'the brain controls thyrotropin secretion', and became an established fact in 1969 when it was declared that 'TRF(H) is Pyro-Glu-His-Pro-NH2'. Latour and Woolgar argue that TRF(H) has meaning and significance according to the context in which it is used: it has a different significance for medical doctors, for endocrinologists, for researchers and graduate students who use it as a tool in setting up bioassays, for a group of specialists who have spent their entire professional career studying it and for whom TRH represents a subfield. But outside this network TRH does not exist. The history of TRH involves many values and choices, including the funding for the project and the crucial moments during the project when it was about to be cut off; choice of strategy involved in the decision to obtain the chemical structure; the imposition of the fourteen criteria which had to be accepted before the existence of a new releasing factor could be accepted; the personalities of the two rival groups in the field and the dispute between them over priority; the dispute over the name of the substance (TRF to TRH); the doubts over the peptidic nature of TRF; and finally, the use of mass spectrometry, which introduced an ontological change in the research and put an end to the dispute. But does all this mean that there is no 'real TRF' just waiting to be discovered, no matter how socially constructed the process of discovery? Latour and Woolgar point out that the transformation of statement into fact is reversible: that is, reality can also be deconstructed.

TRF may yet turn out to be an artefact. For example, no arguments have yet been advanced which are accepted as proof that TRF is present in the body as Pyro-Gly-Ori-NH2 in 'physiological significant' amounts. Although it is accepted that synthetic Pyro-Glu-His-Pro-NH2 is active in assays, it has not yet been possible to measure it in the body. The negative findings of attempts to establish the physiological significance of TRF have thus far been attributed to the insensitivity of the assays being used rather than to the possibility that TRF is an artefact. But some further slight change in context may yet favour the selection of an alternative interpretation and the realization of this latter possibility.[19]

So, the inevitable conclusion: reality cannot be used to explain why a statement becomes a fact, since it is only after a fact has been constructed that the effect of reality is obtained.

Not surprisingly, *Laboratory Life* caused a sensation. But the controversy surrounding it only made the self-consciously Kuhnian constructionists even more important to the interdisciplinary field of science studies (sometimes called science and technology studies, which existed previously as science, technology and society studies). In Alan Sokal and Jean Bricmont's *Intellectual Impostures* (1997), the follow-up book to Sokal's hoax, the strongest venom is reserved for Latour and other constructionists. Sokal and Bricmont are much kinder to another school of sociology of science which does not go as far as the constructionists – namely, the 'Strong Programme' of the Edinburgh School.

The 'Strong Programme' began at Edinburgh University in the late 1960s. Its seminal statement was published just three years before Latour and Woolgar's book. In *Knowledge and Social Imagery* (1976), David Bloor, one of the founders of the programme, stated that the aim of the Strong Programme was both to demonstrate that scientific knowledge is amenable to sociological inquiry and to evolve a methodology for sociology of scientific knowledge (SSK). It had four basic elements. Firstly, the Edinburgh School argued, the goal of SSK is to discover the conditions which bring about states of knowledge; these conditions can be economic, political, and social as well as psychological. Secondly, SSK had to be impartial in its selection of what is studied: giving equal emphasis to true and false knowledge, success and failures of science as well as rational and irrational inquiries. Thirdly, there should be consistency or 'symmetry' in the explanations sought for selected instances of scientific knowledge – one could not use, for example, a sociological cause to explain a 'false' belief and a rationalist cause to explain a 'true' belief. Fourthly, the models of explanation of SSK had to be applicable to sociology itself. One of the concerns of the Edinburgh School was to make scientists more receptive to the concerns of social scientists and to sensitise them to the various social and cultural environments in which their work was embedded. This initiative was part of the general attempt to address what C. P. Snow had called the 'two cultures' problem that first embroiled British intellectuals in the 1950s, when 'technocrats' started to replace Oxbridge humanists in the civil service.

In its early phases, the Strong Programme was seen as truly radical and subverting of science, and generated controversy and fire from the defenders of science in equal measure. However, the Strong Programme accepts, unlike the constructionists, the existence of an unproblematic reality that is successfully explored in science, and in spite of its radical critique, is fundamentally with the positivist camp. The programme's conservative character is eminently displayed in *Scientific Knowledge* (1996), an updating of the programme's position by David Bloor and Barry Barnes. The argument now is that it is not so much the observations in science that are 'theory-laden' but rather the reports of the observations. How an observation is reported depends on the tradition within which a scientist is working. The interpretation of an observation involves bringing to bear the resources of a tradition. So two scientists working in different (scientific) traditions may observe the same thing but report and interpret the same results in different ways. Moreover, according to the Edinburgh School, theories themselves are not fixed in time; nor can they be identified with a set of fixed statements. Association of theories with the names of Great Scientists – 'Newton's Theory', 'Einstein's Theory' – creates this illusion. It is better to think of scientific theories as evolving institutions: a detailed examination of 'Mendel's Theory' shows us how many twists and turns it has taken since Mendel first formulated it.

The Edinburgh School has always argued that 'experience' and 'reality' are actually 'out there'. Realism should not be opposed but illuminated by sociological inquiry. The new conceptual tool they offer for elucidating realism is 'sociological finitism', which is a way of looking at how 'words' are connected to the 'world'. Finitism suggests that all scientific terms and concepts are open-ended – 'no specification or template or algorithm fully formed in the present' is 'capable of fixing the future correct use of the term' – and emphasises the conventional character and sociological interest of scientific classifications. Finitism also suggests that the boundaries between scientific disciplines, as well as the demarcation between what is science and what is not, are liable to change with changing situations. It is thus conceivable that a shift in our conception of what is science may lead to incorporation of what is currently dismissed as non-science into science: astrology, acupuncture, parapsychology, etc. But the Strong Programme does not acknowledge any 'social interest' in science outside of science itself –

there is no sense of larger social forces operating on science beyond those that the scientists themselves witness. So the Strong Programme is not so much a critique of science but, as the Edinburgh School admits, a 'part of the project of science itself'; puzzle-solving within normal sociological science.

This assimilationist attitude is in sharp contrast to the feminist approach to science. For almost half a century, feminists have been suggesting that science discriminates against women. But this discrimination is not only a question of management of science. It is something inherent in science itself. On one level, it is simply the content of science that appals many women – not many are interested in pursuing a career that links science intricately to military and weapons research, or torturing of animals or manufacturing machines that put one's sisters out of work. But the feminist analysis goes much deeper. As Sandra Harding, author of the influential *The Science Question in Feminism* (1986) put it: science is inherently 'androcentric'. Consider, for example, traditional evolutionary theories that tell us that the roots of some human behaviour are to be found in the history of human evolution. The origins of Western, middle-class social life, where men go out to do what a man's got to do, and women tend the babies and look after the kitchen, are to be found in the bonding of 'man-the-hunter'; in the early phases of evolution, women were the gatherers and men went out to bring in the beef. That theory is based on the discovery of chipped stones that are said to provide evidence for the male invention of tools for use in the hunting and preparation of animals. However, if you look at the same stones with different cultural perceptions, say one where women are seen as the main providers of the group – and we know that such cultures exist even today – you can argue that these stones were used by women to kill animals, cut corpses, dig up roots, break down seed pods, or hammer and soften tough roots to prepare them for consumption. You now have a totally different hypothesis; and the course of the whole evolutionary theory changes. Other developments in science, such as the rise of IQ tests, behavioural conditioning, foetal research, sociobiology, can be analysed with similar logic. Gender bias thus emerges in the way basic questions are asked in science. The kind of data that is gathered and appealed to as evidence for different types of questions enhances this bias further. Feminist scholarship of science, which is truly monumental both in terms of quality and quantity, has analysed almost every branch of science.[20] It has shown that the focus on quantitative measures, analysis of variation, impersonal and excessively abstract conceptual schemes, is both a distinctively masculine tendency and also one that serves to hide its own gendered character. And it has revealed that the prioritising of mathematics and abstract thought, standards of objectivity, the construction of scientific method and the instrumental nature of scientific rationality, are all based on the notion of ideal masculinity.

Would a fair representation of women in science change anything? To begin with, it would have obvious economic advantages. Knowledge-based economies, in dire need of trained scientists, cannot afford to squander half of their scientific potential. There is also the argument that more women in science would open up science to a wider range of material and social problems. For example, the problems of the Third World would receive greater emphasis and more research support. But the feminist scholars are arguing for something more. They suggest that women would introduce a shift away from conventional scientific method and objectivity to what Harding calls 'strong objectivity' and Hilary Rose calls 'responsible rationality'. Strong objectivity requires natural scientists to take perspectives of the 'outsiders' – the social scientists, the environmentalists, the housewife, the non-Western cultures – into their description and explanations of the subject of scientific inquiry. The argument here is not that feminist notions of science should be recognised as legitimate and desirable alongside the conventional practice of science. Nor that anti-sexist concepts, theories, methods and interpretations should be regarded as scientifically equal. Nor even that more women should be trained and recruited to work alongside colleagues and within institutional norms and practices that are obviously discriminatory, so that women have to become men in order to practice science. The argument is that having a fairer representation of women in science will not actually solve the problem; science will continue to be discriminatory. Only a fundamental transformation of concepts, methods and interpretations in science will produce real change. Feminist scholars are asking for nothing less than a reorientation of the logic of scientific discovery.

Post-colonial Studies of Science

Kuhn's *Structure* is about how science works in one civilisation: the Western civilisation. But Kuhn can be used to argue that other civilisations, with their own paradigms, would have different practices, and indeed kinds, of science. This idea was first floated in 1978 by the Muslim scholar Hossein Nasr in his highly influential work, *The Encounter of Man and Nature*. Nasr argued that what makes Western science distinctively Western is its conception of nature. The idea that nature is there only for the benefit of man (*sic*) and as Bacon put it, has to be 'tortured' to reveal its secrets, is totally alien to most non-Western cultures. Islam and China, for example, do not look at nature as an object. In Islam, nature is a sacred trust that has to be nurtured and studied with due respect and appreciation. In Chinese tradition, nature is seen as a self-governing web of relationships with a weaver, with which humans interfere at their own peril. Similarly, Western ideas of the universe and time are culturally-based. The Western idea of the universe as 'a great empire, ruled by a divine logos' owes more to centralised royal authority in Europe than to any universal notion – it is totally incomprehensible to the Chinese and Indians. Similarly, while Western science sees time as linear, other cultures view it as cyclic (as in Hinduism) or as a tapestry weaving the present with eternal time in the

Hereafter (as in Islam). While modern science operates on the basis of either/or Aristotelian logic (X is either A or non-A), in Hinduism logic can be fourfold or even sevenfold. The fourfold Hindu logic (with the extra forms: X is neither A nor non-A; nor both A and non-A; nor neither A nor non-A) is both a symbolic logic and a logic of cognition. It can achieve a precise and unambiguous formulation of universal statements without using the 'for all' formula. Thus the metaphysical assumptions underpinning modern science make it specifically Western in its main characteristics. A science that is based on different notions of nature, universe, time and logic would therefore be a totally different enterprise from Western science.[21]

The conventional (Western) history of science, however, does not recognise different types of civilisational or cultural sciences. It has represented Western science as the apex of science, and maintained its monopoly in four basic ways. First, it denied the achievements of non-Western cultures and civilisations as real science, dismissing them as superstition, myth and folklore. Second, the histories of non-Western sciences were largely written out of the general history of science. Third, it rewrote the history of the origins of European civilisation to make it self-generating. Many notable scientists, Newton in the late seventeenth century and Kelvin in the late nineteenth century among them, were involved in creating and disseminating the revisionist history of the origins of modern European civilisation and the creation of the Aryan model. This model introduced the idea that Greek culture was predominantly European, and that Africans and Semites had nothing to do with the creation of the classical Greek civilisation. But the identification of Greek culture as European is questionable on several grounds. For one thing, the idea of 'Europe', and the social relations such an idea made possible, came centuries later – some would date it to Charlemagne's achievements, others to the fifteenth century. (Greece and Rome were civilisations of the Mediterranean.) Moreover, it was Islam that introduced Greece to Europe; and due to the spread of Islam, the diverse cultures of Africa and Asia can also claim Greek culture as their legacy. Fourth, through conquest and colonisation, Europe appropriated the sciences of other civilisations, suppressed the knowledge of their origins, and recycled them as Western. We know that many scientific traditions were appropriated and fully integrated into Western sciences without acknowledgement.

Thus the pre-Colombian agriculture that provided potatoes and many other food crops was absorbed into European agricultural practice and science. Mathematical achievements from Arabic and Indian cultures provide another example. Francis Bacon's three great inventions that made modern Europe – printing, gunpowder and the magnetic compass – are now all admitted to have come from China. Knowledge of local geographies, geologies, botany, zoology, classification schemes, medicines, pharmacologies, agriculture and navigational techniques were provided by the knowledge traditions of non-Europeans.

Within the last thirty years, we have seen the emergence of a new brand of scholars of science. Mostly based in the 'Third World', these post-colonial scholars, engaged in what is known as post-colonial science and technology studies, set out to reclaim the history of non-Western science and expose the Eurocentrism of Western science. It began with empirical work in the history of Islamic, Indian and Latin American sciences, to show the sheer scope and breadth of these sciences. But post-colonial scholarship goes much further. First, it seeks to establish the connection between colonialism, including neo-colonialism, and the progress of

Western science. For example, in his several books, Deepak Kumar,[22] the Indian historian and philosopher of science, has sought to demonstrate that British colonialism in India played a major part in the way that European science developed. The British needed better navigation, so they built observatories, funded astronomers and kept systematic records of their voyages. The first European sciences to be established in India were, not surprisingly, geography and botany. Throughout the Raj, European science progressed primarily because of the military, economic and political demands of the British, and not because of the purported greater rationality of science or the alleged commitment of scientists to the pursuit of disinterested truths. Second, post-colonial scholarship of science seeks to re-establish the practice of Islamic, Indian or Chinese science in contemporary times. There is, for example, a whole discourse of contemporary Islamic science[23] devoted to exploring how a science based on the Islamic notions of nature, unity of knowledge and values, public interest and so on, could be shaped. A similar discourse on Indian science has also emerged in the last decade.[24]

Critique of Kuhn

Kuhn's work has enabled wide-ranging critiques of science to develop since the mid-1960s. Indeed, Kuhn has been seen, and is often presented, as subversive of science. But Kuhn's radical credentials are more apparent than real. We can argue that any analysis of scientific activity is subversive of science, and is often recognised and resented as such. In as much as *Structure* invites analysis of science, it is a radical text. After all, it is not accidental that the teaching of science is, as Kuhn said, as dogmatic as theology, or that the history of science purveyed in textbooks is like Orwell's *Nineteen Eighty-four*. It is part of the schizoid self-consciousness of science to which Kuhn reacted but which he did not analyse. That is, there is openness and much debate at the research front, but certitude and dogmatism in teaching and propaganda. This duality is a product of the embattled tradition of science, fighting against theology as the entrenched source of Truth; but it is also very convenient as a means of control of science's own territory. For to admit uncertainty means to share legitimacy and power – and who would do that willingly? Certainly, Kuhn was concerned to maintain the legitimacy of science: his real interest was in showing that all the key processes of science – including its messy discovery phase – could be explained in terms of science's self-organising principles. While Kuhn tried to expose the problematic nature of science as a historical process, he was very concerned to preserve its internal purity and faith in its organising principles. Those who called for reforms in science admitted that much of the inspiration for change came from outside science itself. Kuhn wanted to deny this and show that science itself was capable of internal reform and change.

If science could reform itself, through revolutions, what need was there for outside interference with science? The arguments of *Structure* could thus be used successfully to exclude the forces of challenge and contamination, such as religion, ethics and technology. Thus, Kuhn became instrumental in marginalising all those critics of science who had argued against science's increasing involvement with the military-industrial complex. *Structure* contributed to the maintenance of an internal/external dichotomy in science. This distinction became pervasive particularly in the teaching of the history of science, and in the cultivation of appropriate, safe historiographical attitudes in historians of science; and it eventually became a general strategy for doing research in the history of science. Moreover, the very features in Kuhn's account that enabled him to distance the nature of science from its most destructive contemporary manifestations – namely, the omission of science's technological, economic and cultural dimensions – were used by social scientists to think that they could reinvent themselves as 'real scientists'.

Kuhn was particularly concerned to preserve the public face of science. Whatever the 'internal' problems and 'truth' of science, the public's belief in science as Good and True had to be defended, for the social consequences of not doing so could be devastating. Indeed, the public's loss of faith in science could even lead to the end of civilisation as we know it. This doctrine of 'double truth' in fact has a long history, going right back to Plato and his reservation about the public display of critical reason following the fall of Athens. Steve Fuller has used the term 'embushelment' to describe the fear of publicly contradicting received opinion because of its potentially destabilising social consequences. Embushelment is responsible for the notion that significant cultural artefacts are doubly encoded, with one message intended to appease the masses by reinforcing their prejudices and the other meant only for élite inquirers who are mentally prepared to assimilate a strongly counter-intuitive truth. Kuhn's own embushelment was a product of his background in the academic milieu of Harvard – which after promoting the Bomb went on to play a leading role in the protracted Cold War – and his personal history with Conant. All of this led him to conclude that in a fickle, bipolar world, the autonomy of science had to be defended and protected from marauding outsiders such as Marxists and New Agers. Kuhn's concern to persevere with the public image of science led him, in later life, to deny that he was a Kuhnian!

Despite *Structure*'s revolutionary pretence, it has been used largely to reinforce the hackneyed old images of science. In his brilliant biography, *Thomas Kuhn: A Philosophical History for Our Times* (2000), Steve Fuller shows that most of the legitimatory uses of *Structure* have been entirely conservative, from Daniel Bell's use of Kuhn's theory to reinforce the role of disciplines over inter-disciplinary research in the besieged universities of the late 1960s, to the more recent invocation of Kuhn by Francis Fukuyama to support the view that science's autonomous development has enabled it to be the motor of global wealth production. The influence of *Structure* on the philosophy of science did nothing to enhance its critical attitude towards science. Nowadays, philosophers are content to attend to the norms implicit in the particular sciences they study, which are presumed to proceed in a normatively desirable fashion. As to be expected in a Kuhnified world, much of this change in philosophical orientation, Fuller shows, has been accompanied by a rewriting of the field's own history. After the end of positivism, Kuhn provided a new focus for philosophical debate, so it was easier to marginalise Ravetz, Feyerabend, the radical tradition of science, and much of the post-colonial critique of Western science. *Structure*'s redefinition of the philosophical agenda meant that constructionism became the focus of science criticism at the expense of the argumentative and rhetorical sides of scientific inquiry. Terms such as 'reason' and 'rationality' went through constant revisions, so that now radical criticism of science has come to be associated with irrationalism.

Beyond the Science Wars: Post-Normal Science

What is at stake in the science wars? Is it simply the destructive influence on science of 'postmodernists' and other scholars of science studies as suggested by

Sokal and Bricmont? Is it about errors and mathematical howlers made by constructionists? Or is it about the power and prestige of science?

Science wars clearly amount to much more than an academic hoax and the exposure of mathematical ignorance in social scientists and cultural theorists. Sokal's hoax proves what many scholars already suspected: cultural studies has become quite meaningless, and anyone can get away with anything in the name of postmodern criticism. There is, however, no evidence to suggest that these scholars have had any real effect on the financial establishment and the public support enjoyed by science. But if science war is about anything, it is largely about the power and authority of science. The fury of the scientific community stems from its recognition that the traditional legitimacy of science is eroding; and the authority of science has haemorrhaged beyond repair. But science war tells us little about why this is happening. For that, we have to look at science itself and how it has changed since the First World War. To a very large extent, science war has become irrelevant – discussion has now moved on to detailed critique of case studies, as can be seen from post-Sokal debates.[25] It will, no doubt, continue to generate debate and controversy. But the fate of science lies elsewhere.

Science is simply not what realists and idealists claim it to be. Its ideological and value-laden character has been exposed beyond doubt. But it is not simply a question of how political realities of power, sources of funding, the choice of problems, the criteria through which problems are chosen, as well as prejudice and value systems, influence even the 'purest' science. Nor that value-commitments, realised in the choice of 'confident limits' of statistical inference, can be found at the heart of scientific method. Nor that most of the assumptions of science are those of the European civilisation. It is more an issue of how science is now associated with uncertainties and risks. A great deal of contemporary science is no longer normal science in Kuhnian terms. As can be seen from a string of recent controversies, from the BSE affair in Britain to the issues of genetically modified foods, science cannot deliver hard and fast answers to a host of contemporary issues. The old paradigm of science which provided certainty and assurance is no longer valid. Science has moved into a post-normal phase in which, to use the words of Ravetz and Funtowicz, 'facts are uncertain, values in dispute, stakes high, and decisions urgent'.[26] The conventional, old-paradigm normal science may still be valid in situations with low levels

63

of uncertainty and risk, but it is not suitable when either decision stakes or system uncertainties – as, for example, in the case of genetic engineering or human cloning – are high. The moral panic of scientists is rooted in this reality – the shifting paradigm that has changed the context of science and brought the uncertainties inherent in complex systems to the fore.

There is no get-out clause here: scientists have to confront this new reality. They may deny that blind faith in science, and the trust and confidence it inspired in the public, have now gone forever; but this would not change the public's perception of science. Post-normal science requires science to expand its boundaries to include different validation processes, perspectives, and types of knowledge. In particular, it requires the gap between scientific expertise and public concerns to be bridged. Thus, post-normal science becomes a dialogue among all the stakeholders in a problem, from scientists them-selves to social scientists, journalists, activists and housewives, regardless of their formal qualifications or affiliations. In post-normal science, the qualita-tive assessment of scientific work cannot be left to scientists alone – for in the face of acute uncertainties and unfathomable risks, they are amateurs too. Hence 'there must be an *extended peer community*, and they will use *extended facts*, which include even anecdotal evidence and statistics gathered by a community. Thus the extension of the traditional elements of scientific practice, facts, and participants creates the element of a new sort of practice. This is the essential novelty in post-normal science.'[27] It inevitably leads to a democratisation of science. It doesn't hand over research work to untrained personnel; rather it brings science out of the laboratory and into public debate where all can take part in discussing its social, political and cultural ramifications.

As Ravetz and Funtowicz point out:

Some people are uncomfortable with the idea that this new sort of practice is science. But science has continuously evolved in the past, and it will evolve further in responding to the changing needs of humanity . . . The traditional problem-solving strategies of science, the philosophical reflections on them, and the institutional, social, and educational contexts need to be enriched to solve the problems that our science-based industrial civilisation has created. To experience discomfort at the discovery of the uncertainties inherent in science is a mark of nostalgia for a secure and simple world that will never return.[28]

Notes

1. The conference proceedings were published as Paul Gross, Norman Levitt and Martin Lewis, eds., *The Flight from Science and Reason*, New York: New York Academy of Sciences, 1996.

2. Paul Gross and Norman Levitt, *Higher Superstition: The Academic Left and Its Quarrels with Science*, Baltimore: Johns Hopkins University Press, 1994, p. 2.

3. Andrew Ross, 'Introduction', *Social Text* 46–7, 1996, pp. 1–13 (p. 8).

4. Ibid., p. 9.

5. *Social Text* 46–7, 1996, pp. 217–52. In the interest of constructionist objectivity, I should mention that Sokal cites this humble author in his bibliography!

6. 'A physicist experiments with cultural studies', *Lingua Franca*, May/June 1996, pp. 62–4.

7. N. Bukharin *et al.*, *Science at the Crossroads*, London: Frank Cass, 1971.

8. J.D. Bernal, *Science in History*, London: Pelican, 1954, volume 1, p. 1.

9. Thomas S. Kuhn, *The Structure of Scientific Revolutions*, Chicago: University of Chicago Press, 1962, p. v.

10. Ibid., p. 10.

11. Ibid., pp. 5–6.

12. Ibid., pp. 2–3.

13. Ibid., p. 15.

14. Ibid., p. 10.

15. Ibid., p. 20.

16. Barry Barnes, *T. S. Kuhn and Social Science*, London: Macmillan, 1982, p. 9.

17. Ian Hacking, ed., *Scientific Revolutions*, Oxford: Oxford University Press, 1981, pp. 1–2.

18. Jerome R. Ravetz, *Scientific Knowledge and Its Social Problems*, New Brunswick: Transaction Publishers, second edition, 1996, p. xi.

19. Bruno Latour and Steve Woolgar, *Laboratory Life: Social Construction of Scientific Facts*, Princeton, New Jersey: Princeton University Press, 1986, second edition; this quotation is from the excellent review by John Stewart, 'Facts as commodities', *Radical Science Journal*, No. 12, 1982, pp. 129–37, p. 132.

20. See, for example, Donna Haraway, *Primate Visions: Gender, Race and Nature in the World of Modern Science*, New York: Routledge, 1989; Hilary Rose, *Love, Power and Knowledge*, Oxford: Polity Press, 1994; Margaret Wertheim, *Pythagoras' Trousers*, London: Fourth Estate, 1997.

21. On non-Western alternatives to and in science, see the numerous essays in Ziauddin Sardar, ed., *The Revenge of Athena: Science, Exploitation and the Third World*, London: Mansell, 1988.

22. See Deepak Kumar, *Science and the Raj*, Delhi: Oxford University Press, 1995; and Deepak Kumar, ed.,

Science and Empire, Delhi: Anamika Prakashan, 1991.

23. See Ziauddin Sardar, ed., *The Touch of Midas: Science, Values and Environment in Islam and the West,* Manchester: Manchester University Press, 1984; Ziauddin Sardar, *Explorations in Islamic Science,* London: Mansell, 1985; and the special issue on Islamic science of *Social Epistemology,* 10 (3–4), July–December 1996, pp. 253–8, ed. Ahmad Bouzid.

24. See, for example, Susantha Goonatilake, 'The voyages of discovery and the loss and re-discovery of "Others" knowledge', *Impact of Science on Society,* 167, 1992, pp. 241–64; and the *Proceedings of the Congress on Traditional Sciences and Technologies of India,* 28 November–3 December 1993, Bombay: Indian Institute of Technology, 1993.

25. The realist attack continues in Noretta Koertge, ed., *A House Built on Sand: Exposing Postmodernist Myths About Science,* New York: Oxford University Press, 1998; the powerful constructionists' defence is presented by Thomas Gieryn, *Cultural Boundaries of Science: Credibility on the Line,* Chicago: Chicago University Press, 1999. Hand to hand combat between the two sides is presented in *Social Studies of Science,* 29 (2), April 1999, pp. 163–315, which is devoted to science wars.

26. S.O. Funtowicz and J.R. Ravetz, 'Three Types of Risk Assessment and the Emergence of Post-Normal Science', in S. Krimsky and D. Golding, eds., *Social Theories of Risk,* Westport, Connecticut: Praeger, 1992,

pp. 251–73, p. 254.

27. Ibid.

28. Ibid., p. 255. For further discussion on post-normal science see Jerome Ravetz, ed., 'Post-Normal Science', Special Issue *Futures*, 31 (7), September 1999.

Further Reading

Adas, Michael, *Machines as the Measure of Men: Science, Technology and Ideologies of Western Dominance*, Ithaca, New York: Cornell University Press, 1989.

Barnes, Barry, *T.S. Kuhn and Social Science*, London: Macmillan, 1982.

Barnes, Barry; Bloor, David; and Henry, John, *Scientific Knowledge: A Sociological Analysis*, London: Athlone, 1996.

Barr, Jean, and Birke, Lynda, *Common Science? Women, Science and Knowledge*, Bloomington: Indiana University Press, 1998.

Fuller, Steve, *Thomas Kuhn: A Philosophical History for Our Times*, Chicago: University of Chicago Press, 2000.

Fuller, Steve, *Science*, Buckingham: Open University Press, 1997.

Fuller, Steve, *The Governance of Science*, Buckingham: Open University Press, 2000.

Funtowicz, Silvio, and Ravetz, Jerome R., *Uncertainty and Quality in Science for Policy*, Dordrecht: Kluwer, 1990.

Gibbons, Michael, *et al.*, *The New Production of Knowledge*, London: Sage, 1994.

Harding, Sandra, *Is Science Multi-Cultural: Postcolonialism, Feminisms and Epistemologies*, Bloomington: Indiana University Press, 1998.

Harding, Sandra, ed., *The Racial Economy of Science*, Bloomington: Indiana University Press, 1993.

Jacob, Margaret, ed., *The Politics of Western Science: 1640–1990*, New Jersey: Humanities Press, 1992.

Kuhn, Thomas S., *The Structure of Scientific Revolutions*, Chicago: University of Chicago Press, 1962.

Longino, Helen, *Science as Social Knowledge*, Princeton, New Jersey: Princeton University Press, 1990.

Midgley, Mary, *Science as Salvation*, London: Routledge, 1992.

Ravetz, Jerome R., *Scientific Knowledge and Its Social Problems*, New Brunswick: Transaction Publishers, second edition, 1996.

Ravetz, Jerome R., *The Merger of Knowledge with Power*, London: Mansell, 1990.

Sardar, Ziauddin, ed., *The Revenge of Athena: Science, Exploitation and the Third World*, London: Mansell, 1988.

Sardar, Ziauddin, *Explorations in Islamic Science*, London: Mansell, 1985.

Selin, Helaine, ed., *Encyclopaedia of the History of Science, Technology and Medicine in Non-Western Cultures*, Dordrecht: Kluwer, 1997.

Sokal, Alan, and Bricmont, Jean, *Intellectual Impostures*, London: Profile Books, 1998.

Woolgar, Steve, *Science: The Very Idea*, London: Tavistock, 1988.

Key Ideas

Constructionism is the idea that scientific knowledge is merely 'constructed', and that this is done through social processes which are no more rule-governed or edifying than those of other areas of life. Once one is aware of the provisional and contested character of science on the research front, the idea of a simple discovery, or successive 'unveiling' of the reality of Nature becomes implausible. But thoroughgoing constructionism must cope with its own implausibility: are scientific knowledge and technical control purely arbitrary, just the result of power struggles? Can we explain their success solely in that way? Or must there be some reality testing somewhere? But for constructionism to address these questions means losing its rhetorical force.

Normal science is the dominant practice of science, involving the use of standard scientific techniques and procedures for gathering, sorting, processing, and applying information. In normal science, the vast majority of scientists are engaged in puzzle-solving that involves working on minute scientific problems within established theories. A process of peer reviewing controls both the boundaries of accepted work as well as its quality. This method of quality control is appropriate only for situations with high levels of certainty and low levels of risks.

Paradigm is the key technical term in Kuhn's philosophy of science. Basically, a paradigm is a way of looking at things: a set of shared assumptions, beliefs, dogmas, conventions, theories. Kuhn is generally concerned with how **paradigm shifts** occur as science develops.

Post-colonial science studies is based on the assumption that science, as it is practised today, is an ethno-science, reflecting the metaphysical assumptions and incorporating the historical trajectory of Western civilisation. The study of science from post-colonial perspectives, which has emerged over the last three decades, looks at such notions as 'the Dark Ages', 'the scientific revolution' and linear 'scientific progress' with radical scepticism. Post-colonial science studies aims at: (1) reclaiming the history of non-Western sciences, technologies and medicines such as Islamic, Indian and Chinese sciences, as well as indigenous knowledge of other cultures, through empirical and historical scholarship; (2) evolving a contemporary discourse of the nature, style and scope of non-Western sciences, technologies and medicines; and (3) developing science policies that recognise and promote the practices of non-Western sciences, technologies and medicines, as well as practices of indigenous knowledge systems.

Post-normal science is the emerging mode of science. All the old debates about scientific knowledge are rendered irrelevant by the new challenges to science – the plethora of emerging problems in which 'facts are uncertain, values in dispute, stakes high and decisions urgent'. These are the cases in which either scientific uncertainty and/or ethical uncertainties or conflicts are extreme. In such circumstances, seeking Truth is a diversion; results are characterised by their quality, which itself is a contextual and recursive attribute. Here Kuhn's normal science, the puzzle-solving 'applied science' of routine work, gives way to post-normal science. In post-normal science, the quality assurance of scientific inputs to policy processes requires an 'extended peer community', consisting not just of the experts but of all stakeholders. And they can bring in their 'extended facts', not just traditional results, but also local knowledges, community surveys, leaked documents and investigative journalism. Only through post-normal science can scientific endeavour recover from the loss of morale and commitment that started with the Bomb, was accelerated by Kuhn and constructionism, and is now rampant under the capture of science by globalisation.

Scientific realism asserts that science actually discovers reality, and that scientific facts are not merely 'invented' or constructed. Ever since the seventeenth century there had been a problem that the reality portrayed by science is not the same as that which we perceive with our senses; thus the earth goes around the sun, and matter is composed of tiny invisible particles. But starting with Kuhn, it was realised that 'science' at any one time can state theories that are later proved incorrect; thus in the nineteenth century scientists believed that heat was a substance called 'caloric' and that light travelled in a rigid medium called the 'luminiferous ether'. Scientific realism had been protected by historical accounts showing that people who held or defended such erroneous views were bad scientists, or were not following Scientific Method. Kuhn destroyed this argument, and suggested that scientific belief has an element of the 'arbitrary'.

Strong Programme was developed in the mid-1970s at the Science Studies Unit, University of Edinburgh. The 'Edinburgh School' sees the sociology of scientific knowledge (SSK) as part of the project of science itself, an attempt to understand science in the idiom of science. The Strong Programme aims to evolve a specific methodology for SSK that is impartial, consistent, and applicable both to sociology as well as science. It is based on the assumption that 'experience' and 'reality' are actually 'out there' and that scientific realism can be illuminated by socio-logical inquiry. The Edinburgh School honours science by imitation; but it has been criticised for professing a 'voodoo' sense of social causality of science that relies on standards of necessary and sufficient conditions so stringent that not even physics could satisfy them.

Other titles available in the Postmodern Encounters series from Icon/Totem

Derrida and the End of History
Stuart Sim
ISBN 1 84046 094 6
UK £2.99   USA $7.95

What does it mean to proclaim 'the end of history', as several thinkers have done in recent years? Francis Fukuyama, the American political theorist, created a considerable stir in *The End of History and the Last Man* (1992) by claiming that the fall of communism and the triumph of free market liberalism brought an 'end of history' as we know it. Prominent among his critics has been the French philosopher Jacques Derrida, whose *Specters of Marx* (1993) deconstructed the concept of 'the end of history' as an ideological confidence trick, in an effort to salvage the unfinished and ongoing project of democracy.

Derrida and the End of History places Derrida's claim within the context of a wider tradition of 'endist' thought. Derrida's critique of endism is highlighted as one of his most valuable contributions to the postmodern cultural debate – as well as being the most accessible entry to *deconstruction*, the controversial philosophical movement founded by him.

Stuart Sim is Professor of English Studies at the University of Sunderland. The author of several works on critical and cultural theory, he edited *The Icon Critical Dictionary of Postmodern Thought* (1998).

Foucault and Queer Theory
Tamsin Spargo
ISBN 1 84046 092 X
UK £2.99   USA $7.95

Michel Foucault is the most gossiped-about celebrity of
French poststructuralist theory. The homophobic insult
'queer' is now proudly reclaimed by some who once
called themselves lesbian or gay. What is the connection
between the two?

This is a postmodern encounter between Foucault's
theories of sexuality, power and discourse and the
current key exponents of queer thinking who have
adopted, revised and criticised Foucault. Our
understanding of gender, identity, sexuality and cultural
politics will be radically altered in this meeting of
transgressive figures.

Foucault and Queer Theory excels as a brief introduction
to Foucault's compelling ideas and the development of
queer culture with its own outspoken views on
heteronormativity, sado-masochism, performativity,
transgender, the end of gender, liberation-versus-
difference, late capitalism and the impact of AIDS on
theories and practices.

Tamsin Spargo worked as an actor before taking up her
current position as Senior Lecturer in Literary and
Historical Studies at Liverpool John Moores University.
She writes on religious writing, critical and cultural
theory and desire.

Nietzsche and Postmodernism
Dave Robinson
ISBN 1 84046 093 8
UK £2.99   USA $7.95

Friedrich Nietzsche (1844–1900) has exerted a huge influence on 20th century philosophy and literature – an influence that looks set to continue into the 21st century. Nietzsche questioned what it means for us to live in our modern world. He was an 'anti-philosopher' who expressed grave reservations about the reliability and extent of human knowledge. His radical scepticism disturbs our deepest-held beliefs and values. For these reasons, Nietzsche casts a 'long shadow' on the complex cultural and philosophical phenomenon we now call 'postmodernism'.

Nietzsche and Postmodernism explains the key ideas of this 'Anti-Christ' philosopher. It then provides a clear account of the central themes of postmodernist thought exemplified by such thinkers as Derrida, Foucault, Lyotard and Rorty, and concludes by asking if Nietzsche can justifiably be called the first great postmodernist.

Dave Robinson has taught philosophy for many years. He is the author of Icon/Totem's introductory guides to Philosophy, Ethics and Descartes. He thinks that Nietzsche is a postmodernist, but he's not sure.

Baudrillard and the Millennium
Christopher Horrocks
ISBN 1 84046 091 1
UK £2.99   USA $7.95

'In a sense, we do not believe in the Year 2000', says French thinker Jean Baudrillard. Still more disturbing is his claim that the millennium might not take place. Baudrillard's analysis of 'Y2K' reveals a repentant culture intent on storing, mourning and laundering its past, and a world from which even the possibility of the 'end of history' has vanished. Yet behind this bleak vision of integrated reality, Baudrillard identifies enigmatic possibilities and perhaps a final ironic twist.

Baudrillard and the Millennium confronts the strategies of this major cultural analyst's encounter with the greatest non-event of the postmodern age, and accounts for the critical censure of Baudrillard's enterprise. Key topics, such as natural catastrophes, the body, 'victim culture', identity and Internet viruses, are discussed in reference to the development of Jean Baudrillard's millenarian thought from the 1980s to the threshold of the Year 2000 – from simulation to disappearance.

Christopher Horrocks is Senior Lecturer in Art History at Kingston University in Surrey. His publications include *Introducing Baudrillard* and *Introducing Foucault*, both published by Icon/Totem. He lives in Tulse Hill, in the south of London.